ETHICAL HACKING

The Complete Beginners Guide to Basic Security and Penetration Testing

(Networking Basics and Ethical Hacking for Newbies)

Timothy Carbajal

Published by Jennifer Windy

Timothy Carbajal

All Rights Reserved

Ethical Hacking: The Complete Beginners Guide to Basic Security and Penetration Testing (Networking Basics and Ethical Hacking for Newbies)

ISBN 978-1-989965-95-5

All rights reserved. No part of this guide may be reproduced in any form without permission in writing from the publisher except in the case of brief quotations embodied in critical articles or reviews.

Legal & Disclaimer

The information contained in this book is not designed to replace or take the place of any form of medicine or professional medical advice. The information in this book has been provided for educational and entertainment purposes only.

The information contained in this book has been compiled from sources deemed reliable, and it is accurate to the best of the Author's knowledge; however, the Author cannot guarantee its accuracy and validity and cannot be held liable for any errors or omissions. Changes are periodically made to this book. You must consult your doctor or get professional medical advice before using any of the suggested remedies, techniques, or information in this book.

Upon using the information contained in this book, you agree to hold harmless the Author from and against any damages, costs, and expenses, including any legal fees potentially resulting from the application of any of the information provided by this guide. This disclaimer applies to any damages or injury caused by the use and application, whether directly or indirectly, of any advice or information presented, whether for breach of contract, tort, negligence, personal injury, criminal intent, or under any other cause of action.

You agree to accept all risks of using the information presented inside this book. You need to consult a professional medical practitioner in order to ensure you are both able and healthy enough to participate in this program.

Table of Contents

Introduction

All the best techniques and tricks on how to learn ethical hacking, penetration testing, network infiltration, and password cracking are in this book! Only the best techniques are in this book.

Chapter 1: Understanding Ethical Hacking

Ethical Hacking is an act of performing and testing security on IT infrastructure with proper authorization from a company or organization. A person performing ethical hacking is known as ethical hacker or computer security expert. An ethical hacker will use latest hacking tools and social engineering techniques to identify vulnerabilities on IT infrastructure.

Overall the ethical hacking provides risk assessment about the security of IT infrastructure for a company or organization information systems. These risk assessment information will provide the level of security that can be exploited by a hacker.

On the other hand, hacker is a person who breaks into IT infrastructure or computer networks without any authorization. Hackers mostly hack for profit or motivated by challenge. These exploitation can cause financial lost, legal impart and trust towards the organization.

THE IMPORTANCE OF IT SECURITY

Nowadays all the companies or organizations are using and depending on IT infrastructure, computer networks and computer systems to operate their core businesses. Most companies store their client information in the server database systems. A good hacker will easily break into customer database if weak passwords are utilized on the server.

Definitely this will cause heavy financial losses to the company. Mostly these hacked incidents will not be reported in the media in detail because it will spoil the company's reputation.

Moreover shopping and bill payments are performed online these days. Therefore client's credit card information must be protected at all cost. One of the most famous method to gain client's credit card information is by performing spoofing.

Objective of spoofing is to fool the user into thinking that they are connected to the trusted website.

Most attacks are implemented utilizing emails these days. A good example would be the Love Letter worm attacks performed. Millions of computers have been attacked and made changes to the users' system itself. The Love Letter worms are received using email attachments.

IT security is crucial to the organization and individual computer users. Individual computer users must make sure they have installed the latest antivirus and anti-spyware in their computers. Companies must ensure they have engaged a computer security expert to look into their computer network security issues.

ETHICAL HACKING PROCEDURES AND STRATEGIES

The first step in performing ethical hacking is to understand a hacker's process.

There are basically five main steps and processes of hacking:

1) Gaining targeted information

2) Probing vulnerabilities for exploitation

3) Gaining access to the targeted system

4) Maintaining access on targeted system

5) Covering the tracks on targeted system.

The targeted system is mostly referring to the machine to be hacked. It can represent a server or computer or any electronic devices. The hacker will perform the five steps mentioned above to gain control or steal information or stop the machine services. Each steps above may take a few months to achieve the desired goal.

An ethical hacker will perform the same steps above to further understand the weaknesses of the targeted system. Once the weaknesses are identified, the ethical hacker will take steps for countermeasure to avoid further exploitation on the targeted system.

Chapter 2: Ethical Hacking And Criminal Hacking

With the **beginner's guide**, you learned that there are two different kinds of hackers. Ethical hackers, and criminal hackers. Each hacker is going to use techniques that are going to be similar to get into the system that they want access to. However, there are some major differences between the two besides just the definition of their names.

Hackers that are ethical

Ethical hackers can be referred to as white hat hackers because they are using their hacking skills to get into a system for good. They are not getting into the system to harm a company or an individual. Instead, they are going to find all the weaknesses in the system and therefore they are going to be helping people to add extra security so that others cannot get into the system.

As discussed in the Introduction, it does not matter if you are hacking on your own, or if you are hacking for a company, you have to know programming languages.

To become an ethical hacker, you are also first going to have some sort of experience in the IT field. Military service can also count as IT experience because joining the military offers IT classes if you enroll in a military specialty that is related to this field. Not just that, but having military service is going to look favorable on your resume.

You should get the certification that is for A+. Having other certifications and experience in the technical field are going to be a major plus for you. The higher up in the position that you can get, the more experience you are going to have obviously because you are going to be proving your skills over and over again and demonstrating that you can handle more responsibility and technical challenges.

There are security certifications that you should look into getting as well that will then help you get a position with information security. In the position that you obtain in information security, you should strive to get through the testing that is offered for penetration of systems while you are getting used to using the tools that are needed for hacking.

The next thing that you are going to want to do is work towards the certification that is going to label you a Certified Ethical Hacker that the International Council of Electronic Commerce Consultants offers.

At the point in time that you have gotten all of the certifications and experience necessary, you can now annotate that you are an ethical hacker on your resume and begin looking for a job in the field.

You are not just going to need to have technical skills to be a hacker, you also need to have skills with people, the ability to manipulate programs, run programs, be good at solving problems, and have a work ethic that is going to stand out among others that may be trying to get jobs as ethical hackers as well.

To remain an ethical hacker, you need to be sure that you are not falling into any hacking activities that are going to be considered illegal.

Criminal hacking

Just like with ethical hacking, criminal hackers are known as black hat hackers. They are the ones that are going to use their hacking skills to harm others.

As a criminal hacker, you are going to use the same tools that are going to be used for ethical hacking, the only difference is that you are not going to get the education or certificates that will label you as a hacker that is out there to help others.

Criminal hackers are going to be using their skills for fraud, theft, vandalism, and terrorism. All these crimes are going to be done on a computer and will most likely involve the internet in one way or another. And, criminal hackers are not going to limit themselves to hacking into companies or the government, they will also try and get into personal computers as well because they are going after any information that is going to help them and harm the person that they have targeted.

The biggest thing to remember about criminal hacking is that it is going to be considered a federal offense and is going to be taken seriously inside of the States. When you are looking at the law, you are going to realize that hacking is going to be defined as gaining access to a system without having the proper permission or going over that permission to get a hold of access that has been restricted from your view. This information can be any number of files that are saved onto a computer be it government, business, or individual.

The tools that are used for hacking will be used in order to complete these tasks. Trojan horses are going to be used to appear as a program that is legitimate, but instead it is going to give a criminal hacker access to the system as a backdoor, but the user of the system is never going to know that they put a virus on their computer until it is too late.

Programs such as Sniffer are going to be used to get passwords so that access can be granted into various platforms without having to break through the security that may be in place. Viruses and other spyware is going to be another way that hackers are going to have access to any system that they want.

Criminal hackers work on the fact that not everyone pays attention to what is going on with their system. Viruses are going to be placed in various programs or emails that are going to work their way into a system and therefore will not be detected until the hacker has the information that they are looking for and by that time, it is too late.

Criminal hackers do not always get away with the crimes that they commit. Many actually end up getting caught and end up spending time behind bars as well as having a fine that they have to pay for what they have done.

Believe it or not, there are some very famous names that have contributed to the advancement of technology that first started out as a criminal hacker but then turned their lives around and helped with the technological world that we enjoy today.

Punishments

- If you are found hacking in India which means that you are tampering with a computer or destroying files you are going to be fined up to 20000 rupes which is $294.85but you will also be placed in prison for up to three years at least. You may end up getting a longer sentence depending on what you were doing and what the situation is.

- It is also in India that if you hack into a website to tamper with it then you are going to also get up to three years in prison, but your fine is going to be 50000 rupees. That is $ 737.12.

- In the Netherlands, hacking is known as having any work that is going to intrude upon the automated work that is going to go against the law. Any intrusion is going to be using log ins that you stole, sending out false signals, or even breaking past security measures. You will end up with a year in prison and a fine with a mark on your record that is a felony in the fourth category.

- The United States is going to forbid any use of a computer that is not authorized if it is protected. Most of the protected computers in the United States are those that are used for the government or financial institutions.

You do not have to just break into these computers, you can use a computer to interfere with foreign communication whether that computer be located here in the United States or somewhere else. You are tampering with government relations.

The punishment may be a year in a federal prison or a fine but that fine is not going to be above $5,000.

Chapter 3: How To Hack Computer Systems And Networks

According to IT experts, the process of hacking has 5 distinct stages. All hackers (i.e. white hats, black hats, and gray hats) follow these steps. Thus, you have to understand each stage completely if you want to be a great hacker (regardless of the "color" you are aiming for). These stages are:

Reconnaissance – In this stage, the hacker gathers information about his/her target. This is divided into two types:

Passive Reconnaissance – Here, you will gather information without the knowledge of your target. This form of reconnaissance can be as plain as observing an office to determine what time employees enter or exit it. However, information gathering for hacking purposes is often conducted in front of a computer.

You can gather information quickly and easily if you will use the internet. Actually, hackers usually run an online search about the company or organization they are going to attack. You can utilize the power of search engines (e.g. Google, Yahoo!, Bing, etc.) to get the data you need.

You may also use the technique called "sniffing the network" to get technical information such as hidden networks or servers, naming conventions, and IP addresses. This method is similar to office monitoring: the hacker observes the flow of information to determine when transactions occur and where the pieces of data are sent to.

Active Reconnaissance – In this type, you will probe the computer network to find services, IP addresses, and individual hosts linked to it. Active reconnaissance employs aggressive techniques. Thus, it involves higher risks of being detected than the passive type.

This way of gathering information can generate important data (e.g. the security measures being used). However, it can also increase the chances of being caught.

Both types of reconnaissance can result to the collection of confidential data. For instance, it's often easy to discover the kind of web server and OS (operating system) the target organization is using. You can use this data to find the vulnerabilities of your target.

Scanning – In this stage, you will use the information you have gathered to analyze your target. Here are some of the tools used by hackers in the hacking stage:

Dialer

Port scanner

ICMP (Internet Control Message Protocol) scanner

Network mapper

Ping sweep

SNMP (Simple Network Management Protocol) sweeper

Vulnerability scanner

Hackers search for the following pieces of information:

User accounts

Installed software

Computer names

IP addresses

Operating System

Getting Access – This is where the actual hacking takes place. The weaknesses discovered during the previous stages are now used to access the target. The attack can be sent to the target system using a LAN (local area network), local access to a computer, the internet, or offline. Here are some examples: session hijacking, denial of service, and buffer overflows.

This stage is also called "owning the system" because once the target has been hacked; the attackers can control the system as they wish.

Maintaining Access – After establishing the connection, the hacker needs to maintain that access for future attacks and exploitations. In some cases, hackers fortify the target using rootkits, backdoors, and Trojans. They do this to secure an exclusive access to the target and prevent security personnel or other hackers from discovering the vulnerabilities being used.

Once the system is hacked, the hacker can use it to launch other attacks. This strategy is called the "zombie system" – the system is forced to hack other targets.

Covering the Tracks – Once the hacker has gained and secured access, he/she can cover his/her tracks to:

Prevent detection by IT personnel

Remove evidence of their hacking activities

Avoid legal action

Continue using the hacked system.

In general, hackers want to eliminate all traces of their attack like log files and IDS (intrusion detection system) alarms. Here are some of the things they do to cover their tracks:

Modifying the log files

Steganography

Employing a tunneling protocol

The Technology Used by Hackers

Nowadays, there are many tools and methods that you can use to locate weaknesses, run exploits, and attack targets. After finding the vulnerabilities of a system, you may start your attack and install malware (i.e. malicious software). Here are some types of malware: backdoors, rootkits, and Trojans.

SQL injection and buffer overflows are two other techniques used to access computer networks. In general, these techniques are employed when attacking databases and application servers.

Hacking tools are used to exploit one of these areas:

Applications – Often, applications are not checked for weaknesses: programmers don't test their projects during the Code Writing phase. That means any application may have programming errors that you can take advantage of. Basically, the process of creating an application is driven by features – the programmers need to create the most feature-rich version of their project in the shortest period of time.

Misconfigurations – In some cases, computer systems have erroneous setups. Hackers can take advantage of this by discovering those errors and making sure that they won't be detected or corrected by the security personnel.

Shrink-Wrap Code – Lots of ready-made computer programs have bonus features the average user doesn't know about. You can use these features to breach a network's defenses. For instance,

Microsoft Word has the "macros" feature – hackers can use this to trigger programs inside the application.

Operating Systems – Most system administrators apply the default settings when installing new operating systems. Since the default settings are not designed to prevent network attacks, a hacker may find a vulnerability that can be exploited.

Different Types of Hacking Attacks

Hackers employ a wide range of techniques to overcome the target's defenses. In general, hackers try to specialize in two to four hacking techniques. This allows them to have an in-depth knowledge about the attacking method they want to use. Hacking methods are extremely complex: it would be hard and impractical if you will try to master all of them.

When talking to a client, a hacker asks whether there are particular problem areas like social engineering or wireless networks. This information can help the hacker in customizing his attack.

The following list shows the most popular entry points for hacking attacks:

Local Network – A LAN (Local Area Network) attack simulates a person who wants to gain EXTRA and UNAUTHORIZED access to a local network. Here, the hacker should establish access to the network before he can launch this attack.

WLANs (Wireless LANs) belong to this category and provide an extra entry point for hackers as the radio waves pass through solid structures. Since WLAN signals can be determined and tapped outside the building, attackers don't have to access the computer network physically and execute the hacking procedure. In addition, the growing popularity of WLANs has turned this method into a go-to trick for black hat hackers.

Remote Network – Remote network hacks try to simulate an outsider attacking the system through the internet. The hacker attempts to breach or discover weaknesses in the external defenses of the target (e.g. proxy, router, or firewall vulnerabilities). The internet is considered as the most popular hacking instrument. Because of this, many technology-intensive organizations have fortified their defenses against "online hackers."

Stolen Equipment – Here, the hacker steals a data resource owned by a member of the target organization (e.g. employee). Data such as passwords, usernames, encryption types and security settings can be acquired by stealing a physical device (e.g. laptop, desktop, smartphone, tablet, etc.). This area is usually overlooked by lots of organizations: they focus too much on the electronic side of things.

Once hackers gain access to a device authorized by the security domain, many types of information (e.g. security settings) can be obtained. When a laptop disappears, the owner doesn't report it immediately since he/she thinks that the loss is a personal one. This way of thinking is incorrect. Stolen devices must be reported as soon as possible so that the security personnel can prevent the missing gadget from accessing the network. If the owner retrieves the device, he/she can just contact the system administrator to have it unlocked.

Remote Dial-Up Network – Here, the hacker launches an attack against the target's modem pools. For instance, hackers use war dialing (i.e. the process of dialing repeatedly to discover system vulnerabilities) in violating the defenses of a dial-up network. To prevent this hacking method, lots of organizations have replaced dial-up connections with secure internet connections.

Social Engineering – With this point of attack, the hacker tries to exploit the integrity and security of the network's authorized members. Hackers who use social engineering employ face-to-face or telephone communication to get the information they need. You can use social engineering to obtain passwords, usernames, and other security measures of the target.

Physical Entry – This point of attack refers to the target's physical premises. A hacker who obtains physical access to a network can plant rootkits, Trojans, viruses, or key loggers (a device that can record keystrokes). In addition, the hacker may gather confidential files that are not placed in secure locations. Finally, an intruder who gets physical access to the target's building may install rogue gadgets such as wireless access points. A wireless access point allows a hacker to access the local network even if he/she is in a remote location.

Chapter 4: What Is Hacking?

Machines and networks made of machines and the instructions that make them go -- these things are not without their weaknesses -- and the same is true whether we're talking about a production line, an intranet that exists within one division of a multinational company, or the worldwide conglomeration of computers, computer networks, and software that we call the Internet.

You can put locks on the doors of a factory to protect the machines and the flow of the process. You can shut down the modems and take out the cables that connect the computers in the intranet to each other and only to each other. You can't do that with the Internet -- and that's the reason why we need to learn about cybersecurity. Also known as computer security or IT security, cybersecurity encompasses everything to do with the protection of information systems. Cybersecurity seeks to prevent the theft of information held in these systems; it also seeks to protect these systems themselves from damage or intentional misuse.

In the world of cybersecurity, the person who can find a vulnerability or weakness in a system, or who can get into that vulnerability and exploit it, is called a hacker. There are still those who think that hacking is as simple and as easy as getting into someone's Facebook or Twitter account without their knowledge or consent. Or they think that the be-all and end-all of hacking is in defacing a website to show silly or obscene or provocative messages. But these things are small beer in the grand scheme of the world of hackers.

Before you learn how to become a hacker, you should learn about the different types of hackers, and become familiar with their similarities and differences.

Script Kiddie

These hackers use programs, tools, and scripts created by other hackers, instead of creating their own. They don't tend to know how systems work, but will happily use already available tools to hack.

White Hat Hacker

Otherwise known as ethical hackers. White Hat Hackers are the good guys, and they do what they do in order to keep the rest of us safe. By locating flaws in information systems and doing their best to fix them, they improve cybersecurity as a whole. They tend to work for large organizations.

Black Hat Hacker

These are the unethical hackers who hack for malicious purposes. They steal customer data or money, infect a system with malware, or make information systems do things that they're not supposed to do.

Grey Hat Hacker

These hackers are on the fence, so to speak. They don't hack for malicious purposes but will still break into an information system just to show that they have the chops to do it, or just to show that there are flaws in that system. If they do work with a company, they might present a solution to the flaws that they found only AFTER they have hacked into that company's information system.

HACKTIVISTS

The word "hacktivist" is a portmanteau of the words "hacker" and "activist". They break into information systems for the sake of protesting against injustice and for the sake of working towards social justice -- or sometimes just to perform some wild publicity stunts. Hacktivists have been appearing more and more often in the headlines as of late, with the most famous of them being Anonymous.

How to Become a Hacker

The road to becoming a good hacker is long and complicated, and one of the first things to learn before taking the first step on that road is the proper programming language to use. Every website and system is built using one or more computer languages, and in order to hack these sites or systems successfully, you must first understand those languages. The following are the languages that you'll need to understand for each particular purpose.

Web Hacking

If you want to be able to hack websites you'll need to read, code, and understand the following computer languages, all of which play vital parts in the display and functioning of different web-based content.

HTML -- standing for Hypertext Markup Language, this is the standard markup language that is used in creating and modifying web pages.

CSS -- standing for Cascading Style Sheets, this is a style sheet language that is used to define the physical appearance / presentation of a document that has been written in a markup language. As such, it is used hand-in-hand with HTML.

JavaScript -- used in both online and offline forms, it is used to create, support, and display everything from web pages to PDFs.

PHP -- a server-side scripting language used mostly for designing web pages, but it is now also used to create other kinds of computer programs.

SQL -- standing for Structured Query Language, this is used to manage information that is coded into a database, or that is in the form of a data stream.

Writing Exploits

In addition, a hacker with knowledge of the computer languages Ruby and Python will have a huge advantage, as they are both used in writing exploits.

But what is an exploit? An exploit, in the context of cybersecurity, takes advantage of a known or a discovered bug or vulnerability in an information system. It could take the form of a piece of software, an amount of data, or even a sequence of commands. These allow a hacker to get further into an information system than might have been originally intended.

• Ruby -- a relatively programmer-friendly language that's designed to be easy to learn, but it can yield massively complex results. This object-oriented programming language can and has been used in writing exploits, as it has a great deal of flexibility in its syntax and variables. The Metasploit framework, which we will discuss further on in this book, is built on Ruby.

• Python -- another widely-used programming language that supports the creation of programs that are easy to read, no matter the size of the program or the intended functions. It is the core language for creating hacking tools and writing exploits.

Reverse Engineering

There are many ways of looking into a piece of software in order to learn how it's built and how it works -- and in reverse engineering, the hacker works backwards from the published product. The hacker looks at the software and what it does, and deduces the code, the functionalities, and the process flow that might produce the observed results. This might be considered by some to be a low-level form of coding, though it can actually yield very advanced results, especially in cases where the source code/s can no longer be obtained, or where the source code is not easy to find or modify.

Anyone who learns how to reverse-engineer code will be able to break down, understand, and modify many different kinds of programs, as well as many different forms of hardware. A working knowledge of assembly languages will come in handy. There are many different assembly languages, each specific to the architecture of a particular computer system.

These are just a few of the languages that you need to learn in order to become a hacker. Now we are going to look at a way to find most every vulnerability and exploit there is to find.

At the end of the book, you can find a list of resources where you'll find more information on the languages and procedures described here.

Finding Exploits and Vulnerabilities

While there are many tools that can be used in order to investigate the various vulnerabilities of information systems, in this book we will zero in on the SecurityFocus database.

It's a rare thing indeed to find the exploits that you need during your first run at hacking into an information system. Don't rely on blind luck: use your head and use the tools that are already available to you.

First: do some research. What will work on your target, and what won't? Take a look at the operating system that the target is using, and once you've pinpointed that, then it will be easier to look for open ports and exploitable servers. From there, you can determine the best way to compromise the target without detection and that is not always going to be an obvious solution.

You will likely want to put in a little work in order to find the right vulnerabilities to exploit in your target information system -- and then to find the corresponding tools that can work on those vulnerabilities. This tutorial is all about finding those vulnerabilities and exploits.

Step 1

Using a browser that you can trust, head to the URL www.securityfocus.com. The database contains quite a few tools that you can use to search for vulnerabilities. You can run searches in a number of ways including the Common Vulnerability and Exploit number, also known as the CVE number. This number is assigned by the MITRE Corporation, a non-profit organization funded by a section of the US Homeland Security department.

Step 2

The CVE database is full of vulnerabilities. Most every vulnerability that has been found is included here, including those vulnerabilities that the software publishers would prefer to keep hidden. Let's take Adobe for an example. The last few years have not been kind to Adobe thanks to software that has been poorly designed, resulting in releases that are absolutely jam-packed with vulnerabilities. Just about every single computer has Adobe Reader and/or Adobe Flash installed, which leaves that computer and any networks it might be connected to open to an attack. For this section, we are going to focus on Adobe Flash Player.

Step 3

In SecurityFocus, click on the Vendors tab and select Flash Player from the menu. Click on Submit and you will be presented with a list of vulnerabilities that exist in Adobe Flash Player. Information on each vulnerability exists on that list, as do different ways of exploiting those vulnerabilities. For example, one of these vulnerabilities will allow you to install a rootkit or listener on any system running Flash Player. These programs will give you access to that system as though you were an actual registered user on that system.

Step 4

Having found the vulnerabilities in your targeted information system, the next step to take involves finding the actual exploits that might be relevant to your objectives. The remote code execution vulnerabilities are likely to be present on virtually every computer system. A vulnerability is a weakness in the system that can be exploited, but that doesn't mean that someone has managed to get in and perform the actual exploit.

While you're going to need a few advanced skills in programming in order to develop your own exploits, it's easy as long you're talented.

So, to find an exploit for the vulnerability, click the Exploit tab in the SecurityFocus window. In the results you'll see the exploits that have already been developed for the particular vulnerability you chose.

If the vulnerability is brand-new, there will be no exploits. Develop your skills in programming using the languages that were listed in the previous sections, and you can use those skills to be the first person to come up with an exploit for a new vulnerability.

This chapter covered how to find vulnerabilities and how to find the exploits for those vulnerabilities. With this information you can go on to use the exploits to get into a system through a vulnerability.

Chapter 5: Countermeasures

In this section, we will now discuss some countermeasures for use against the clarification techniques discussed above. First, some general countermeasures are presented, followed by special techniques against passive enlightenment, active enlightenment, and social engineering.

General

Common countermeasures include applicable security policy, removal of standard files such as the server software, and the software version number and changing the default settings, such as the default user name and password. It is also important to educate users on secure passwording and handling sensitive information. It is also absolutely necessary to regularly update the software used and to keep track of reports about "security gaps" in the software used.

Passive Enlightenment/Foot-printing

As a countermeasure to Internet research and Google hacking, it is advisable to check the contents of your own website and remove unneeded published information. If this is not possible, it is important to hide sensitive files or directories because Google Crawlers index all files residing on a web server. By creating a robots.txt file containing control statements for the various crawlers, the behavior of the search engine crawlers can be influenced. For example, by specifying in robots.txt, crawlers can be banned from entering certain directories and tracking links. Specifying META tags in the HTML header allows the crawlers to index the pages found in the search engine cache. Setting up password-protected areas using htaccess also helps hide information from crawlers. Sufficient protection of a page can be checked by Google search queries. In doing so, Google Hacking methods, which are also used by attackers, are applied to the site to be protected in order to find unneeded published information. [Long05]

On the other hand, queries by WHOIS and DNS databases, which were also used by attackers, cannot be prevented. This is because the WHOIS entries are required by law and the DNS entries are required to call a domain. This information is thus freely available, and since the query is about databases that are managed by "third parties ", it is unclear to a potential attacker who has requested these entries.

Active Enlightenment/Fingerprinting

The port scan of a system cannot be prevented, but it can be ensured that no unnecessary ports are open. Port scans and thus preparation for attacks can be detected early on if the server logs are analyzed on a regular basis. If such seizure preparations are detected, it helps to regain your own safety precautions. Check and adjust as needed, such as updating the firewall settings to other software or software with known security gaps. Another active option is, with intent, to open wrong ports to swap potential attackers.

OS fingerprinting cannot be prevented, but it can be detected by the server logs. As an active countermeasure to Nmap, IP Personality, a kernel module, which can swap a different system by changing flags in the TCP header.

Social Engineering

It is important to raise awareness of the existence of social engineering and the progression of a social engineering attack. Here a training of the own coworkers helps to handle security-relevant data and particularly for handling social engineering ring. Employees should be trained to unambiguously identify the identity of the counterpart before sensitive data is released. Badly prepared social engineering attacks can be identified by detailed inquiries.

Chapter 6: Hacking

Hacking is the act of trying to penetrate a system using different techniques and tools. In this book, the main focus will be on ethical hacking, and it entails scanning for vulnerabilities within a network while also sealing any loopholes that may be present so that malicious users cannot be able to exploit the network and gain access to sensitive pieces of information.

The term ethical hacking is misused in some instances, and people may also misunderstand the term. Penetration testers and security auditors are classified as ethical hackers. They will focus on finding the vulnerabilities within the system, and they do not also harbor any malicious motives. For an ethical hacker to carry out penetration tests and security audits, they always seek permission from the owner of the network. In this chapter, the main focus will be on the introduction to hacking and what it entails. Also, we will investigate why the hacking process is important.

Many people are conversant with the term malicious users and hackers; the mentioned terminologies are quite confusing and people may be unable to differentiate between both terms. There are also many people who have suffered under the hands of hackers who had malicious motives. It is good to also learn about the malicious attackers. The terminologies that will be used in this book include:

Hackers - These are individuals who try to compromise sensitive pieces of information as well as computers. They normally have some ill motives. They are always in the form of unauthorized users.

The external attackers will always target different systems and they will focus on compromising it. Each hacker will have different preferences and some will always choose to hack into systems that are well protected. Additionally, some attackers will focus more on elevating their status in their attack chain.

Malicious Users - There are some people within the organization that may harbor some ill motives. The malicious users are in the form of "trusted" individuals. A malicious user will always try to hack into a system that they can easily gain access to. The malicious users will always have different techniques and they use specific tools when launching an attack.

Ethical Hackers - They are the good guys and they always try to check for any vulnerabilities within the network. They ensure that they have sealed all the loopholes present within the system so that unauthorized users cannot be able to take advantage of any loopholes. The security experts will fall into this category.

The Definition of a Hacker

The term hacker has two meanings:

Traditionally, a hacker is an individual who had the ability to interfere with an electronic or software system. Hackers like to explore and also learn more about how different computer systems operate. They will also try to investigate different techniques that they can use to access the system both electronically and mechanically.

Recently, the term hacker has gained a new meaning and it means an individual who is trying to break into a system maliciously and they are also after their own personal gain. The criminals are commonly referred to as crackers since they will crack certain systems with the sole aim of gaining access to sensitive pieces of information. Their main goals differ and some people may be trying to exert some revenge while other attackers may be trying to gain financially. The attackers may steal some sensitive information and also delete & modify some of the information that is present within the system with the sole aim of making sure that some people's lives have become miserable.

The white hat hackers do not like being classified in the same categories as the black hat hackers. If you're not familiar with the hacker types, the white hat hackers are the good guys, and the black hat hackers are the bad guys because they also have malicious intentions. There are also hackers of the gray mask. They are both good and bad; their goals will always differ. All in all, the term hacker will always elicit some negative thoughts. The white hat hackers are researchers and they focus on coming up with different tools that they also share with the public. Some of the tools come in handy in different ways.

Who is a Malicious User?

A malicious user is always more of an inside man. The malicious users may be contractors, employees, or even interns. They will always abuse their privileges since they cannot be suspected easily. According to statistics, there are a large number of inside users that are responsible for the attacks taking place within different organizations. The issue about the presence of inside men is among the major security breaches at the moment.

The main focus is not on the hackers. At the moment, the main security breach is the people who are abusing the privileges that they have been accorded within the company. The malicious users will try to look for any sensitive information within the database. They will also go ahead and try to look for any confidential information that may be stored within the emails so that they can use it for their own personal gain. At times, these malicious users may also delete any sensitive information that has been stored on the servers. Some of these people may not possess a lot of technical knowledge about hacking, but they can cause some irreversible damage.

Most of the security professionals do not like dealing with the malicious users since they do not have exact goals and they do not also possess enough knowledge on how to perform an attack. The users only take advantage of the access that they have to the information systems within the company.

Recognizing How a Malicious User Begets White Hat Hackers

It is good to make sure that your network is safe from malicious users and external attackers. The ethical hackers are more of security professionals and such tasks should be delegated to these professionals. The ethical hackers will perform different tests with the sole aim of discovering the vulnerabilities that are present within the system. They will then seal all the present loopholes.

About Ethical Hacking

We have mentioned that the ethical hackers are the good guys and they usually use different penetration testing techniques. They will use the same tools and techniques that are used by the attackers. The main difference is that they do not have any malicious motives. For an ethical hacker to perform a penetration test, they will first seek permission from the target. After discovering various vulnerabilities, they will then provide some evidence to the owner of the network. They will then devise some techniques that will help to seal all the loopholes so that the network or server cannot be accessed by attackers. The ethical hackers are tasked with handling different security improvements. The ethical hackers can also help to validate the claims of vendors who are tasked with selling different security tools such as firewalls.

When performing ethical hacking tests, you should make sure that you are certified and that means that you should have the necessary credentials. For instance, you may have the title- CEH (Certified Ethical Hacker). The certification is issued by the EC-Council.

An Insight into Auditing and Ethical Hacking

There are many people who do not understand the difference between security auditing and ethical hacking. The difference between both terms is huge. As for security auditing, it is the act of comparing the security policies within a company with the present security policies that have been enforced. The main reason why security auditing is important is because it helps to validate the security controls that have been put in place. The approach is risk-based. Auditing also takes place when reviewing different business processes and, in most instances, such tasks are not technical. Security auditing is also more of a security checklist and such tasks are quite simple.

As for ethical hacking, it involves looking into the vulnerabilities that are present within the network or server within the company. The ethical hacking process will help to validate all the security controls that have not been put in place. Ethical hacking can also be quite technical as compared to security audits. The methodology also differs. If security audits take place regularly, it is also good to implement ethical hacking techniques to ensure that the network cannot be accessed by malicious users.

Policy Considerations

In a company or a business, you may deploy various risk management techniques including ethical hacking. To do so, you must make sure that you have documented all the present security testing policies and they will help to outline the types of ethical hacking that should take place. For instance, the policies may help to showcase the systems that should be tested and they may include the laptops, servers, and even web applications. Also, the policies will outline the number of times that the ethical hacking should take place. Other important pieces of information may include the methodology to be used during the ethical hacking process. In some instances, some organizations also come up with various security standards that will also outline the various tools that should be used during the testing process.

Compliance and Regulatory Concerns

After formulating some internal policies, they will be used to manage the security testing within your organization. You should also keep in mind that various federal laws should also be considered since they will also affect your business directly or indirectly. There are many federal laws that you should consider and they include; Payment Card Industry Data Security Standard (PCI DSS), Health Information Technology for Economic and Clinical Health (HITECH) Act, among others. Always make sure that your ethical hacking tests will comply with some of these laws.

Understanding Why You Should Hack Your Own System

If you want to catch a malicious user or an external attacker, you should try to think like them. The approach used by both white hat hackers is the same one used by the black hat hackers. The major difference is that the motives of the two types of hackers will differ greatly. To effectively carry out ethical hacking, you must be able to understand your enemy accordingly. The number of attackers has been growing gradually and that means that you should make sure that your system does not have any exploitable vulnerabilities. It is advisable to ensure that your systems are safeguarded from attacks by malicious users. After learning the tricks used by the external attackers, you will be able to find the vulnerabilities that are present within your system.

An external attacker will always focus on the weak security practices that are present within the system. They will try to force their way through the encryption protocols, firewalls, and they will also try to crack the passwords using different tools. Some of these security systems will focus more on the vulnerabilities that are termed as high-level. When you attack your own systems, you will be able to learn more about the present vulnerabilities. After that, you can fix all the available vulnerabilities so that the external attackers may not be able to penetrate the system. If you fail to investigate the present weaknesses, it means that your system may not be safe and an external attacker will eventually try to exploit all the present vulnerabilities.

Hackers are always trying to learn more about how they can attack different systems since technology is also evolving rapidly. You should make sure that you can think like the attacker so that you can protect your systems accordingly. Ethical hackers should know more about the activities that they can carry out so that they can stop the efforts of the attackers. They should know what they should look for and how they can use the knowledge and tools that they possess.

It is not paramount to protect your system from every potential attack since you cannot. Since ethical hackers seal the present vulnerabilities, the attackers will realize that the system is not accessible and they will always go back to the drawing board. The external attackers will always devise different techniques and that means that the ethical hacking tests should also be carried out at intervals to make sure that the system is safe.

To stay safe, you should be able to focus on all the present vulnerabilities that are present within the system; nevertheless, you should also know that it is not possible to anticipate all the forthcoming attacks. You should focus more on making sure that the system is not prone to unlike attacks. As an ethical hacker, you should have the following goals:

Prioritizing a certain system so that you may be able to focus on the areas that matter.

Ensure that you have hacked the systems in a manner that is not destructive.

Look for the vulnerabilities and also collect some evidence that will also be presented to the management.

Apply all the results so that you can get rid of the present vulnerabilities within the system.

Understanding the Issues That Your System Is Facing

There is a major difference between knowing that your system is susceptible to attacks and knowing the specific vulnerabilities that may be utilized by the external attackers. In this section, the main focus will be on some of the well-known attacks.

There are many security vulnerabilities, but that does not mean that they are all critical. When you decide to exploit many vulnerabilities at a go, the system may crash. Some of the vulnerabilities that cannot be termed as critical include weak passwords and a server that may be hosted on a wireless network. Although these vulnerabilities are not critical, when an attacker exploits both at a go, there is a high likelihood that the system may crash. Also, some sensitive information may also be disclosed in the process.

Learning About Complexity

Ethical hackers will always scan for vulnerabilities since they may be present; it does not necessarily mean that some vulnerabilities are present. Since technology has also been advancing significantly, the number of attacks that has been taking place has increased gradually. The IT environment has also become more complex with the advent of cloud computing and social media.

Nontechnical Attacks

Some of the major exploits involve manipulating the end-user and the system itself in case there are some vulnerabilities present. The external attackers will always take advantage of the trusting nature of human beings and they will always carry out social engineering attacks in the process. Such an attack is nontechnical and it aims at ensuring that the external attacker can gain access to different pieces of information that will come in handy when executing an attack. More information about attacks will be discussed in the preceding chapters.

Other attacks are normally physical. The external attacker may break into the building and access the computer room and other areas that may contain some sensitive information. During a physical attack, the external attacker will try to gain access to different pieces of information that may have some value; as a result, they may use different techniques such as dumpster diving, which involves going through the trash can looking for disks and other pieces of information that may come in handy when facilitating an attack.

Network Infrastructure Attacks

An attacker can easily attack a network since they are accessible through the internet. The attacker will have to bypass the firewalls among other security protocols regardless of their location within the globe. Some examples of a network being attacked include;

Connecting to a network through a wireless access point that is not secured.

Installing network analyzers that will also capture all the packets that are traveling across the entire network. Some confidential information will also be revealed in the process.

Flooding the entire network with numerous resources and also coming up with a DoS (denial of service) attack for all the legitimate requests.

Exploiting all the weaknesses that are present in different network protocols, including TCP/IP.

Operating System Attacks

Most of the external attackers will always try to hack into the operating systems within an organization. Such an attack is common since the computers within the target organization have operating systems. Additionally, operating systems also have different vulnerabilities. There are many operating systems and some of them are more secure as compared to others. Most of the attackers mostly focus on the operating systems that are used by a considerable population and they include Linux and Windows. Also, the weaknesses of the mentioned operating systems are also well-known. Some of the attacks carried out on operating systems include:

Attacking some of the authentication systems that are in-built.

Exploiting some of the missing patches.

Cracking the weak encryption and passwords.

Breaking the file system security.

The Application of the Attacks

There are many applications that are usually targeted by attackers. Some of the programs that are targeted by attackers include web applications and e-mail servers. The breakdown is as follows;

SMTP (Simple Mail Transfer Protocol) and HTTP (Hypertext Transfer Protocol) applications are prone to attacks since they have been configured with firewalls and other mechanisms that are meant to enforce security.

VoIP (Voice over Internet Protocol) attacks have also become common since there are many businesses that are endorsing such communication techniques.

Unsecured files may also contain some sensitive information and they may also be scattered throughout the office. The database systems also contain many vulnerabilities and external attackers can exploit them.

An Insight into the Ethical Hacking Commandments

The ethical hackers will always carry out specific attacks that are also meant to showcase the vulnerabilities that may be present within the system. Some of these attacks may be against the physical controls or the computer systems. After discovering the present weaknesses, the ethical hacker will then seal all the loopholes that can be exploited by the external attackers. To ensure that the security professional is carrying out an ethical attack, you must ensure that they are conversant with the ethical hacking commandments. If they fail to adhere to these commandments, some bad things may happen in the process. In the instances whereby an ethical attacker fails to adhere to the set commandments, the results are not always positive; on the other hand, when they follow the commandments are followed to the letter, the results are normally positive.

How to Work Ethically

To work ethically, you should make sure that as an ethical hacker you can adhere to various professional standards and principles. Also, an ethical hacker should make sure that they have supported the goals of the company that has gone ahead to procure their services. Also, no hidden agendas should be present. Ethical attackers should make sure that they have presented all their findings regardless of whether they may be positive or negative.

Ethical hackers should be trustworthy and they should not also misuse any information that they possess. The external attackers are the ones that misuse confidential pieces of information. When the external attackers are caught, they are presented before a court of law and they may also be jailed for some time.

Respecting Privacy

As an ethical hacker, you should know that all the information that you have gathered is confidential. During the testing phase to look for the present vulnerabilities, make sure that you have gathered enough evidence to justify that different vulnerabilities are present. There may be some web application flaws and some of the passwords may also be easily identified and that means that the system can be exploited by malicious users. Such information should also be kept in a confidential manner. The ethical hackers should also not go ahead and snoop into the private lives of some of the employees since they also have nothing to gain from such an instance. Ethical hackers should also involve some individuals in the testing process so that they can also try and gain the trust of their clients and that means that they can gain access to more tasks that involve carrying out ethical attacks to identify vulnerabilities.

During the attack process, make sure that the systems will not crash. When carrying out an ethical attack, you can also try to hack into your own system and in the process, the system may also crash if you do not possess specific pieces of information that will guide you on how to successfully launch an attack. For starters, the system may crash in an instance whereby there is poor planning. At times, the individuals carrying out the tests may also lack knowledge about the tools that they are using when carrying out an ethical attack.

In some instances, an ethical attacker may also initiate a DoS (Denial of Service) attack without their knowledge when performing an ethical attack. When numerous tests are also carried out simultaneously, some of the data may be corrupted, there might be a system lockup, and a reboot. Some of these issues come about when testing some applications and also websites.

During the attack process, you may also create a system lockout accidentally when using some of the vulnerability scanners since some external attackers may also be carrying out some social engineering and that means that they can also change the passwords without even realizing the consequences of their actions. As an ethical hacker, always tread with caution and they should also make sure that they have fully understood the present DoS weaknesses.

The vulnerability scanners will always control the number of tests that can be carried out simultaneously. The settings also come in handy when carrying out tests on specific systems during the usual business hours. The tests may take a lot of time, but the most important point to note is that discovering the present vulnerabilities is the main agenda.

An Insight into the Ethical Hacking Process

When handling any tasks involving security auditing, you must first come up with a plan. If you fail to come up with a plan, there is a high likelihood that you will fail. A plan will ensure that you have worked strategically and you will also be able to handle issues in a tactical manner. Also, as an ethical hacker, you should also be determined. When you have a plan, you are also bound to succeed.

Some organizations may opt to hire a hacker who has "reformed." In such an instance, it is good to be cautious. In this book, we will also discuss some of the pros and cons that are associated with hiring hackers.

How to Formulate a Plan

For starters, you must ensure that as an ethical hacker, you are certified. Also, make sure that everything that you are doing is known to the general public, specifically the decision-makers within the firm since they are the key players. Ensure that you have also received sponsorship from the client and proceed with the ethical hacking process in an authorized manner.

The authorization process may be in the form of an internal memo that may be issued by one of the managers so that you can go ahead and test the systems for any vulnerabilities. When testing numerous systems for different clients, it is good to make sure that you have signed a contract first. Even if the ethical testing process is called off, your efforts as an ethical hacker will not be wasted. Make sure that you have also documented the entire process so that you can also present some evidence in case some allegations are issued against you.

You must make sure that you fully understand what you are doing since one mistake can lead to the system crashing and that will not be good at all. Make sure that your plan is well detailed and that does not mean that you need to come up with numerous testing procedures. Make sure that the scope is well defined and it should include the following pieces of information;

The specific systems that will be tested- when you are selecting a system, you should make sure that you have started by testing some of the critical systems since they may be vulnerable and highly targeted by attackers. For example, you may start by testing numerous Operating System passwords and web applications. Despite being an ethical hacker, you may try to carry out a social engineering attack since it will shed more light on how to prevent the physical attacks by malicious users.

Evaluate the risks involved- make sure that you have a contingency plan for something may go wrong unexpectedly. For example, you may be assessed the firewall and you may take it down unexpectedly. In the process, there will be system unavailability and the system performance will be reduced in the process. The employees' productivity will also reduce drastically. Also, there will be no data integrity and the ethical hacker's image will be tainted.

Make sure that you have handled the DoS (Denial of Service) attacks and the social engineering attacks carefully.

Ensure that you have investigated the dates that the tests will be performed- start by looking into the tests that have been performed and make sure that you have thought really hard about everything. For instance, are the tests going to be performed during the normal business hours? At what specific time will the tests be performed? Will you involve other parties when coming up with the timing process? Make sure that you have made an informed decision in such an instance.

Some of the consequences that you may suffer are Dos-related and the best approach to such a situation is to carry out an unlimited attack and any of the tasks will be possible at any given moment. When the external attackers are trying to get into the system or network, their scope is not limited and that means that the scope of the ethical hacker should also not be limited; however, it is good to make sure that there are some limitations when it comes to the social engineering, DoS (Denial of Service), and physical security attacks.

Do you want to be detected or not? - When hacking into a system, if you do not want to be detected, you just make sure that you have erased all the digital footprints that you had left behind. In most cases, one of your major goals may entail not being detected. For instance, you may be trying to test a system remotely and you do not want to issue any warning to the users since they will alter their behavior.

As an ethical hacker, you must possess knowledge about the specific systems that you are going to test- ensure that you have extensive knowledge about the systems that you are going to test so that you may also be able to protect the systems.

When you possess enough knowledge about the systems means that the ethical hacking process will progress as planned and you will also encounter minimal challenges. When testing a system for a specific client, make sure that you have dug deeper. In some instances, you will also notice that the IT managers in different organizations are also afraid of the security assessments; as a result, they will also take more time and that means that the assessment will not yield the expected results. As an ethical hacker, make sure that your focus is on the needs of the clients.

Ensure that you have a plan on how to seal vulnerabilities if there any in the system - You may find some security loopholes after starting the testing phase; that does not mean that you should stop the assessment. Make sure that you have progressed with the hacking process so that you may be able to discover all the loopholes that are present within the system. Hack the entire system until you are not able to progress further. If there are no vulnerabilities, it means that your assessment was not thorough. When you uncover some serious information, make sure that you have initiated the discussion with some of the key players within the organization.

The specific deliverables - In this instance, you must make sure that you have come up with the vulnerability scanner reports depending on your findings. Also, make sure that you have described all the countermeasures that are present within the system.

Selecting Tools

When handling any project, you should make sure that you are equipped with the necessary tools. For instance, when carrying out ethical hacking, you should make sure that you are conversant with the tools that are offered by Kali Linux. The operating system has more than 600 tools and you should make sure that you can be able to choose the specific tools that you should use during the ethical hacking process. If you fail to use the right tools, you will not be able to discover the vulnerabilities that are present within the network. The experience that you possess also matters.

Also, make sure that you know the technical and personal limitations. Additionally, some of the vulnerability scanners may also generate some inaccurate information. In some instances, you may need to utilize multiple vulnerability scanners, especially when you are testing different web applications.

Each tool will always focus more on specific tests and there is no tool that can be used to test all vulnerabilities. For instance, a screwdriver has its specific purpose and it cannot be used in some areas. The same case applies to the tools that are used to discover some of the present vulnerabilities. The more tools that you possess, the better. There are both software and hardware tools. Also, the availability of tools will make sure that the hacking process can progress accordingly with minimal hitches.

Make sure that you are using the right tools for different tasks:

When cracking passwords, make sure that you are using cracking tools. Some of the tools that you may use include Proactive Password Auditor and Ophcrack.

When analyzing web applications, you can use NTOSpider and Acunetix. The mentioned tools are more preferable as compared to tools such as Wireshark that are used to analyze networks.

When you are selecting the tools that you are going to use when performing ethical hacking, make sure that you have consulted different security experts. They may issue some subtle advice. Also, search engines such as Google come in handy, especially when you are trying to gain some insight into different tools that you can use when scanning for vulnerabilities within a system. Some tools are available for commercial use and others are available free of charge. The tools that can be used by ethical hackers include:

OmniPeek.

Cain & Abel.

WebInspect.

QualysGuard.

Metasploit.

Ophcrack.

CommView for Wi-Fi.

GFI LANguard.

A discussion about the software and hardware tools that an ethical hacker can use will be initiated in the preceding chapters. We will also discuss more about the actual attacks and how a hacker should plan an attack and execute it in the process. Although there are many tools that can be used during the hacking process, it is evident that each tool has different capabilities.

Since each tool has different capabilities, there are people who have misunderstood some of the tools that are used during the ethical hacking process. The misunderstanding arises as a result of the complexity of some of these tools. When using specific tools, make sure that you fully understand them so that you may avoid some unnecessary confusion. After making sure that you have an in-depth understanding of all of these tools, you will be ready to carry out the attack and the results that you yield will be positive. Some of the ways to ensure that you are knowledgeable about the tools that you are using include;

Make sure that you have read the readme file.

Use the tools in a test environment.

Study all the user guides.

Adequate documentation.

Take part in a training program. Some of the vendors train people on how to use the tools that they have provided.

Make sure that you have come up with a detailed report after finding some vulnerabilities within the system. In the report, make sure that you have included information about how some of these vulnerabilities have been exploited.

Make sure updates are available.

General industry acceptance.

Present a high-level report to the key shareholders within the target organization. The reports are important, especially when carrying out a security audit.

Some of these features will make sure that you have saved some time when carrying out the tests and also formulating the final reports.

Executing the Ethical Hacking Plan

As an ethical hacker, you should make sure that you are persistent. Additionally, you should exercise some caution when carrying out an attack. For instance, some malicious users may be keeping track of your activities and they may use the information that they obtain for their own personal use and that means that the business will be affected adversely.

Make sure that there are no external hackers in the system before you initiate the ethical hacking attacks. Also, make sure that the attack is taking place privately, especially when you are transmitting the results that you have received. It is advisable to ensure that you have encrypted the emails and the files that may contain some sensitive information. Some of the encryption tools that you may use include Pretty Good Privacy. The tools will encrypt the compressed documents that have been stored in the form of a zip file. Alternatively, you can also use similar forms of technology.

Always carry out a reconnaissance mission since it comes in handy when gathering information about the target organization. Ensure that you have adopted the mindset of a malicious attacker and start with the broad view as you narrow your focus progressively:

Search engines come in handy. You can carry out the search in a manual manner. For instance, you can start by searching for the name of the organization and the names of their computer systems.

Eventually, you may try to search for their IP addresses. The most popular search engine is Google and it is a great place to start.

Make sure that you have narrowed your scope and target specific systems within your organization. In some instances, the attacker should start by assessing the physical structures within the organization and the web applications since they will help to reveal a lot of information about the target organization.

Make sure that you have narrowed your focus critically. You can then perform the scans and other tests that are meant to uncover the present vulnerabilities that are present in the system.

Perform an attack while also exploiting the vulnerabilities that you have discovered.

Evaluate the Results

After discovering the vulnerabilities, make sure that you have also assessed the results that you have found. Make sure that you are equipped with the necessary knowledge. Practicing comes in handy, and you will end up garnering more knowledge about the available systems. The evaluation process will also become simpler with time.

Make sure that you have submitted a formal report to the clients and outline all the results and recommendations depending on your findings. Make sure that the parties are kept in the loop so that they can realize that the resources that they have invested were well spent. After the tests are over, you should implement all the recommendations that you had issued so that the system may not be vulnerable to attacks by malicious individuals. If you fail to implement the recommendations, it means that all the resources that the client had invested will go to waste

There are new security vulnerabilities that tend to appear from time to time and that means that the information systems will continue to become complicated. The new vulnerabilities are not usually uncovered and that means that the ethical hacking tests should be carried out regularly at certain intervals. You may end up discovering new vulnerabilities with time and that means that you will be able to ensure that the system is more secure. At any given moment, some changes may take place. For instance, when you update certain applications and also apply patches, some changes will also take place. Make sure that you have planned the tests regularly and make sure that you are also consistent.

Chapter 7: Hacker Lifestyle

Hacking is not just a thing you do, hacking is a lifestyle. This chapter covers what you need to do in order to live the lifestyle.

Mindset

First things first, you've got to get the mindset if a hacker down. Hackers enjoy solving problems. While many people will shy away from problems or would rather live a life without problems, that is not the way of a hacker. A hacker can't stand a life without problems, as hackers love problems. If there is no problem to tackle, if there is nothing to solve, then a hacker feels bored, empty, and without purpose.

A hacker would rather be tackling a problem than doing anything else. The high of solving problems is the hacker's purpose. If a problem seems impossible to solve, all the better. There is nothing that a hacker loves more than tackling the impossible! Because hackers make the impossible possible, that is what a hacker does!

If there is something that can't be penetrated, some coding issue that can't be debugged. The hacker's eyes light up, and the hacker focuses relentlessly until the problem gets solved.

No problem is too great for a hacker, no wall is too high, no piece of code too complex, there is always a way, always another angle that can worked.

This is the mindset of a hacker!

Friends

Hackers have only 5 friends, their computer, their programming books, Stackoverflow.com, their fingerless gloves, and other hackers.

Hackers will go out of their way to do anything for their hacker friends. And when you make some hacker friends, it's not uncommon for them to knock on your window in the middle of the night because they need your help with something. You might be tempted to try to convince yourself that you're just dreaming and go back to sleep. You're not just dreaming, your hacker friend is at the window! So slap yourself awake and go on over and open the window to let your friend in.

Clothes

As a hacker, fashion is not such a big thing. But it's best if you dress the part. You'll need weird looking T-shirts that separate yourself from common people, thin high-tech looking jeans with fashionable rips in them, a good high-tech looking jacket, some fingerless gloves, cool shades that you can wear even at night and still see, and a laptop case with shoulder strap. On your feet, running shoes or running sneakers are best, because you never know when it'll be time to run. Sometimes a hacker just has those moments where running is, well, necessary.

Jewelry

Hackers do not wear any jewelry unless it's high-tech, think touchscreen watches and multifunction vibrating bracelets. A pendant or amulet on a thin chain around the neck would also be acceptable.

Hairstyle

Hackers don't leave their hair a mess. Having a sleek looking hairstyle is important. If you don't have hair, that's fine. But if you do have hair, best if you put some gel in it and push it back or spike it. Or if you have curly hair, then curly hair is fine, just make sure it's well trimmed. If your hair is long, then cut it shorter.

Favorite Color

The favorite color of a hacker is most always black. Blue, red, green, or orange also make acceptable favorite colors. Never yellow, purple or pink.

Conventions

It is important that hackers go to conventions to meet other hackers. Sci-Fi conventions, Tech conventions, Programming conventions, and of course Hacking conventions are all acceptable. Be sure to get out there and go to a convention at least once every few months, and try to find and identify other hackers, just look for the fingerless gloves, and start the conversation by talking about debugging issues or something of that nature.

Chapter 8: Getting The Passwords You Need

The next thing that we will have a look at is how to hack passwords so that you can get onto the websites that you want. Hacking passwords is a great thing to learn and work with because you will be able to get ahold of all the information that you want. Depending on which passwords you will get, they could help you to get into a complete computer system, gain access to bank accounts, and so much more. The right password can give you the keys to anything and everything that you might want.

Now there are a few methods that you can use to get ahold of a password. Some will choose to use what is called a 'brute force' attack, which means that they, with the help of a software program, will just keep trying passwords until one of them works. There are also dictionary attacks that will try passwords based on the words out of a dictionary. Both of these options can work if you have a lot of time to wait around, but it can be hard to wait that long unless the password is simple and short.

Another option that some people like to use, and which can be really successful in getting a password is with the use of a keylogger. This is a program that can keep track of all the keystrokes that the user types in and then sends that information back to the hacker, without the user ever having any idea. You can also add in a screen logger that will send the screenshots back to you as the user is doing things on their computer, and you will easily get all the information that you need.

Some hackers like to go with a method that is known as 'shoulder surfing.' With shoulder surfing, you are basically right behind the user, and you can watch as they type in their password. You will obviously be a little bit further behind them, so they don't notice, but for people who are not good at keeping their information hidden, this can usually work. This method can at least provides you with an idea of how many characters are present inside of the password and can limit the choices that you have.

Going with social engineering is a good option when you really want to gain information on what a certain password is. Hackers like to send out fake emails that look like they are coming from a legitimate company. For example, they may send out an email that looks like it comes from your personal bank. If the user is not paying attention to what is going on, they may click on the link and give up their password, which means that the hacker now has the information that they were looking for.

Vulnerabilities that come with passwords

There are basically two types of vulnerabilities that will come along with your passwords. These two vulnerabilities include user and technical. When we talk about user vulnerabilities, we are talking about weaknesses that come because of things that the user is doing wrong. This could be things like weak passwords or when a company is not willing to add harder and stricter guidelines to keep the system safe from possible infiltrators.

One example that shows user vulnerability is when people will pick out one password and then use it for all of their accounts. They do it because it is easier for the user to remember one password rather than have so many different ones, but if a hacker finds just one of the passwords, the first thing that they will do is assume that this password is used in other places and it will grant them access to all of your accounts. Having different passwords will make it easier for you to keep your information safe and makes it much harder for the hacker to steal your information.

There are also some technical vulnerabilities that you may need to be careful about as well. After the hacker has gone through to see if there are any user vulnerabilities they can exploit, they will switch to working on some technical vulnerabilities. Some of the common technical vulnerabilities that you may deal with include:

The application that you are using will show the password as it is being typed

It will show this on the screen, which makes it easier for someone to see it. Most applications will not automatically do this, but the user is sometimes able to change this setting. Shoulder surfers will then have the chance to take a look and see which passwords are being used.

Databases and programs are designed to store your password on occasion

Sometimes the database does not have the right security on it that it should, and this makes it easy for a hacker to access the system and get the information that they need.

Going with a database that does not rely on encryption to help keep things secure

Without a reliable form of encryption, it is possible that a lot of people can get the information easily even without authorization.

It has encryption, but it is not that good

Many developers feel that their source codes are not widely known so they will not put in the effort to strengthen the system's security. This makes it easier for the hacker to get into the system.

How to complete a password hack

After we have taken some time to talk about why a hacker would like to get ahold of these passwords, it is time to go through and learn how to actually do a password hack. To do this, we are going to use the **pwdump3** tool to make it easier and to help us to get this all done and get a hashed password. We will then use the 'John the Ripper' tool because it is a great one to use whether you are relying on Linux or Windows operating systems and can get passwords from both of these platforms. This excludes Mac, but most systems are Windows or Linux, so this is not entirely an issue.

Now, there is a slightly different method that you will use to hack a password based on whether you are using a Windows system or a Linux system. We will first take a look at the steps that you will need to take to hack into a Windows system:

Go to your computer and then open up the C drive. Create a directory and make sure that you call it 'passwords.'

You need to make sure that your computer has a decompression tool installed. A good option is WinZip. If you don't have a program like this on your computer, you should download and install it.

Now it is time to download and install John the Ripper and **pwdump3**. They need to be extracted into the passwords directory that you make earlier.

Type in the command: "c: passwordspwdump3 > cracked.txt"

The output that you will get will be the Windows Security Accounts Manager password hashes. These will all be captured inside the .txt file.

Now you can type in the command "**c: passwordsjohn cracked.txt**"

This will have the John the Ripper tool placed against all the password hashes and your output will be the user passwords that were cracked.

This method can be easy to work with and is pretty simple, but the process will take you a bit of time, depending on how many people are on the system and how complex their passwords are.

As you can see getting a password off a Windows operating system is not too hard to do and it takes just a few simple steps. The process that you need to complete to get passwords from a Linux system will be a bit different, but if you have ever worked with the Linux operating system in the past, then it won't be too hard. The steps that you can take to hack passwords on the Linux operating system include:

Download all the source files on Linux.

When these files are ready, you should type in the command: [root@local host yourcurrentfilename] #tar –zxf john – 1.7.9.tar.gz

This will extract the program, and it will also help you to create a brand new **/src** directory.

Once the **/src** directory is ready, type in the command "make generic."

Now you can go into the /run directory so type in the command: "**/unshadow/etc/passwd/etc/shadow > cracked.txt**.

From here, the unshadow program will merge the passwords and the shadow files and then will input them into the '.txt' file.

Now you can type in the command: **/john cracked.txt**

This is going to help you to launch the cracking process. This one will take you a bit of time, but you should end up with the same kind of output that you got when using the procedure in Windows.

These two processes show that it does not really take that long to hack through a password on either operating system, as long as the password is not too complicated to start with. This is why all your network computers need to have strong passwords that are difficult to guess so that no one can gain access to your network and cause issues. Setting up requirements for the passwords that make them strong and hard to guess will ensure that the whole network is safe from the clutches of a hacker.

Chapter 9: The Effects Of Hacking

When you hack into a computer, you are opening up the system that you are hacking to multiple effects. As the public sees, many of the effects of hacking are going to be bad because the public doesn't always see that hacking is something that is a good thing.

Just like with anything, there are pros and cons to everything and hacking is truly no different. When you hack someone, you are always leaving them open to the malicious effects of what hacking brings about whether you mean to or not.

When you do hack a computer, you are creating a breach in the computer's security. This is placing the victim's sensitive data and privacy at risk. These hacking activities are usually done in order to gain access to confidential information that one tends to keep on their network such as: social security numbers, bank account data, credit card numbers, and personal photographs. There are a few hackers that tend to use this information to harm the one that they have hacked, then there are others who simply "take" this information in order to prove to their "victim" the major security issues that they have in order to get them to be fixed so that the personal and sensitive information that they have taken cannot be taken again.

Once a computer's security system has been compromised, there is also the possibility of the loss or even manipulation of data. A hacker can go in and delete any sensitive information that has been placed on a network once they have gained access to it. Once your system has been hacked, you're at a great risk for all the data that you have on your computer being lost or manipulated in a way that can and most likely will harm you.

One of the biggest things that everyone associates with hackers is identity theft. Identity theft is when someone who is not authorized takes your identity. Your identity is but not conclusive to your social security number, date of birth, or anything else that would identify you as you. This is usually done with a malicious intent and used for the hacker's personal gain or interest.

When you've been hacked, the hacker can actually track everything that you do on your computer thanks to the advances in technology. Key-logging software is what is used in order to track every keystroke that you make on your computer. Thanks to this software, a hacker can instantly gain access to your passwords, your bank accounts, and anything else that they can use to harm you, all thanks to one little program.

DOS means denial of service attack. This happens when a hacker gets into your network and your computer therefore makes computer resources unavailable to any of the authorized users. It is most often that a DOS will attack a website which will then make the website unavailable for a long period of time. This then causes the users of the website to be inconvenienced as well as hampering with the business of the website.

Along with identity theft, stolen information is a big thing that happens when someone hacks your network for a malicious reason. This can be hazardous to anyone, but most particularly to business' that end up being hacked because then sensitive information that they do not want getting out to the public is then released. Not only that, but email address', client information, so on and so forth can be stolen and compromised.

National security can even be put at risk when it comes to hacking. Hackers who hack into the governments networks then have access to the defense system as well as many other systems that will cause there to be grave consequences on the welfare of the nation. When someone hacks into the government, not only is the nation's security at risk, but so is the well-being of the citizens of the United States.

Another effect of hacking is fraud. Hackers can turn computers into zombies by infecting them with internet enabled computer viruses. These computers are then used for activities that are considered fraud such as spamming and phishing attacks on other networks.

How do you know when you have been hacked? Your computer will most likely decrease in its performance speed. You may also begin to notice files that are not supposed to be there. These files may increase in size as well as be modified without you ever touching them. You may also begin to notice that there are changes in your network settings or even frequent disk crashes.

The only way that you're going to be able to protect your computer is by installing a reliable antivirus software as well as making sure that your firewall is enabled before you begin to connect to the internet. Also make sure that you install the system updates on a regular basis.

Chapter 10: 10 Of The Best Security And Hacking Tools Available

While there are loads of different tools available for you to use, many of them are commercial and won't do exactly what you want them to do – not to mention being pricey in some cases. The following are 10 of the best non-commercial hacking tools that you can make use of, with a brief explanation of what each one does.

Nmap

Most people have heard of this one. Network Mapper is free and it is open source. It is used to explore networks or run security audits and was designed as a way of scanning very large networks quickly – it will work on single host systems very well though. Nmap makes use of raw data packets in highly novel ways to see what hosts are on the network, the services each one is offering, what firewalls and packet filters are in place, the operating system each host is using and loads of other useful information. Nmap will run on almost every type of computer and is available on graphical and console versions. It can be used by skilled pros and complete beginners alike and is highly versatile.

Nessus Remote Security Scanner

Nessus used to be open source but has recently been changed to closed source. It is still free to use though and it works with client-server frameworks. It is one of the most popular scanners and is used by more than 75,000 different organizations across the world. In fact, many of the largest business in the world are saving significant amounts of money by using the program to carry out audits on their most critical applications and devices.

John the Ripper

This is one of the fastest password crackers there is, available for lots of different versions of UNIX, BeOS, DOS, OpenVMS and Win32. The primary purpose of the program is to find UNIX passwords that are weak and highlight them. As well as a number of crypt password types, it also supports Windows NT upwards and Kerberos FS, amongst others.

Nikto

Nikto is open source and is a web server scanner that can carry out comprehensive testing on web servers. It tests for a number of items, including more than 3000 files that are potentially dangerous, version specific issues on more than 200 servers and versions that are fond on more than 600 servers. Plugins and scan tools are frequently updated and can be set to automatically update if needed.

SuperScan

SuperScan is a very powerful pinger resolver and TCP port scanner. SuperScan 4 is the latest update and it makes a very nice alternative to Nmap for Windows systems.

P0f

P0f 2 is an incredibly versatile tool that is used for passive OS fingerprinting. It can be used to identify the operating system on the following:

SYN – machines that connect up to your box

SYN+ACK – machines that you connect to

SST+ - Any machine tha you cannot connect to

Any machine that you can observe communications on

In short, it is able to fingerprint pretty much anything. It does all this by listening, not by actively connecting the target.

Wireshark

Wireshark used to be called Ethereal and it is a TK+ network protocol analyzer, a sniffer if you like, that allow you to catch and browse network frame content. The idea of Wireshark is to make an analyzer that is of commercial quality, for UNIX and to provide features that can't be found in closed source sniffers. Wireshark works on Windows, with a GUI and on LINUX, is incredibly easy to use and is able to reconstruct TCP/IP streams.

Yersinia

This is a network tool that is designed to find weaknesses and take advantage of them in layer 2 protocols. It masquerades as a solid framework of testing deployed networks and systems. At the moment, these network protocols are implemented – Cisco Discovery, Spanning Tree, Dynamic Host Configuration, Dynamic Trunking, Hot Standby Router, Inter-Switch Link, IEEE 802.1g and VLAN Trunking. This is, quite frankly, the very best layer 2 kit there is.

Eraser

Eraser is a Windows tool, one of the most advanced security tools there is for removing sensitive data completely from your hard drive. It does this by overwriting it a number of times with selected patterns. It is free to use and the source code is published under GNU General Public License. It really is one of the best tools for securing your data and for making sure that, if you deleted

Putty

PuTTY is free to use and is an implementation of SSH and Telnet of UNIX and Win32 platforms. It also has an xterm terminal emulator and is a must have piece of kit for anyone who wants to SSH or telnet from Windows without having to use the MS Command line clients provided.

Chapter 11: More Hacking Tools And Techniques

Packet Sniffers

We can use certain software and hardware to intercept and log a network's digital traffic. Such software and hardware are called Packet sniffers. They are also called as packet analysers. These tools can capture and decode the data contained in a packet in its raw form, after which they analyze the obtained information. Some packet sniffers generate traffic of their own, thereby working as reference devices. One of such tools is 'Wireshark'. Hackers use this tool to analyze a network's data traffic. The output resulting from the data analysis is then sent to the hacker. Network administrators so as to identify weak spots also use this tool or vulnerabilities in the network during troubleshoot.

Nmap

Nmap (Network mapper) is a tool, which works by scanning a network or networks for identifying hosts that are available. The following are the tasks performed by Nmap on a network:

Nmap sends certain IP packets to all the host computers on a network. The host computers in turn respond to the sent packets. Nmap captures and examines those responses.

It identifies and gives the list of all the ports that are open on a host computer.

Nmap can identify the operating system of a network.

It can also determine an application's name and version number.

.

Important hacking techniques

Dictionary attack

The dictionary attack involves a computer running all the words of the dictionary one by one, to see if any dictionary word matches with the password under attack. This technique requires software to do all the work, as it is impractical for any hacker to sit and enter each and every word of the dictionary as a password string. Simply put, the software will try to guess the password with the help of a dictionary. If the victim has used a simple dictionary word as his password, it is only matter of seconds till the software cracks the word. This technique is ideal when it comes to cracking passwords that use simple text. But, if the password employs random combination of symbols or numbers, this technique will fall flat.

Hybrid attack

To counter the shortcomings of the dictionary attack, hackers use the hybrid attack for cracking the password. This technique combines the words of the dictionary with some special characters and numbers. These special characters and numbers are prepended or appended to the dictionary word in an effort to uncover the actual password. Also, the symbols and numbers at different positions replace the characters of the word.

. For instance let us suppose the password of the user is “p@s$W0rd789”. In the hybrid attack, the dictionary word ‘password’ is replaced at several positions with the symbols @, $ and numbers to uncover the actual password.

Rainbow table

Passwords are sometimes stored with a hash code attached to them. So, even if you uncover the password using the dictionary and hybrid attacks, you can't break into the system because of the presence of the hash code. Even if you get of hold of the password database, the password needs to be decrypted to its original form. So, you could add a hash code to every dictionary word, before comparing it with the original password. If luck favors you, you might find the correct match.

Brute force attack

If the password is too strong to crack and the hacker is left with no choice, he will still resort to the 'Brute force' attack. The Brute force technique relies on the trail and error method while producing different combinations of passwords. This technique works by combining all the alphabets, digits and symbols in several ways until the correct password string is obtained. So, thousands of test strings will be produced to match against the correct password. Even though the brute force attack consumes a lot of time and CPU resources, it will ultimately crack the password in the end.

Even though the time taken by a brute force attack to crack a password mainly depends upon the speed with which the CPU works and the strength of the password, in the end, the password will be cracked and the account will be hacked.

The advantages and disadvantages of this technique can be summarized as follows:

Advantages:

➢ A hacker can crack any textual password using the Brute force technique, as it involves combining all the possible alphabets, numerals and symbols to produce a test password with every iteration.

Disadvantages:

- ➢ This technique is not time efficient, as it is very time consuming to produce thousands of character combinations.

- ➢ If the computer running the brute force attack is slow, it will take even more time to crack a password.

Note: Brute force attack is different from dictionary attack, as brute force uses every possible character combination to produce the password, while the dictionary attack confines the character combinations to the words of a dictionary. Simply put, brute force attack involves producing strings of characters that don't make sense, while the dictionary attack produces meaningful words.

Chapter 12: What Do I Need To Know About Networks, And How Do They Help Me Hack?

It's incredibly important as a discerning future hacker to know everything that you can know about networking protocols. As we've discussed before, this is because virtually **every** computer is connected through networking services to the internet, which means you can basically access any computer you'd like if you simply know how to.

In other words, the short answer to the question "what do I need to know about networks?" is "everything". There's no shortage of useful information regarding them, and you are never going to hit a point while learning about them that a piece of information could be **not** useful at some point in your hacking career. That is, if you are serious about hacking.

So right now, we're going to talk about the very basics of networking.

Firstly, we have to talk about what networking even entails.

Networking is the manner by which computers talk to one another and communicate essential information. This can occur in multiple ways, and there are numerous different manners in which is does consistently.

The way that these are defined is by **protocols**. Protocols are standards which enable and define the communication and data transfer between two computers. Protocols are, for lack of a better term, the **rules** which define the syntax of the communication between multiple different computers.

Protocols can be implemented by any number of ways, but normally they're implemented by either software or hardware, or a combination thereof.

Why is this important specifically? Simply because these protocols take place on something called **ports**. These ports are where connections, inbound and outbound, can be made.

So what's so special about ports?

Imagine there's a really big building, a skyscraper. And you needed a way in. But every window and door was locked. All you need is a single entry point to get into that skyscraper and get anything you needed in the entire building. The port can provide that entry point. The port is essentially a door left unlocked on the backside of that building.

After getting the target's IP address, you can run a port scan of the machine using the nmap utility. This will enable you to know if they have any open ports that you can connect to their computer by. You can also scan for specific ports in nmap.

So how do you know which ports to look for? Well, blithely, you don't. However, doing a port scan will help you in that endeavor.

It's also important to note that some of the most popular ports in a computer, such as FTP or HTTP, normally have very heavy security. Their only vulnerabilities are generally through exploits which haven't yet been discovered. You could be, in time, the lucky programmer to discover those vulnerabilities and exploit those ports. Then ethically report the issue so that it's no longer a problem, of course. But for right now, look for other ports. A good number of less important ports, such as Telnet on port 23, may be ripe for you to gain entry through, however.

Well that's great and all. Maybe you found an open port? What could you do then? Well, generally, you'd find our what service and operating system are running, then search for a vulnerability.

How do we search for vulnerabilities then? Well, it depends. If you were more experienced or wanted to be more directly connected to exactly what you're doing, you could go to a site like http://exploit-db.com in order to seek out and find exploits for a given port, knowing the version and service running on the port.

However, it's important to note that an open port doesn't necessarily mean a vulnerability exists. It doesn't really mean this at all, actually. All that an open port means is that a connection has been made to the outside world. It's vulnerabilities that allow you access to the host computer.

As a newer user, it may be better to use **metasploit** in order to find the vulnerabilities. As with anything else, this is a highly relative experience and is best learned through experimentation.

Chapter 13: Hacking Mobile Devices

Whether you are wanting to use your phone as a hacking tool to hack into someone else's phone, used as the hacking instrument, or any combination of the two, this is for you.

Use Their Phone as a Keylogger

When someone is sitting at his or her desk, where do you think the phone goes? Typically, right beside the keyboard. This is for convenience, of course. Researchers from Georgia Tech have created a program that can turn someone's cellphone into a keylogger. Now, you might think from that statement that it is using the camera or audio cues in order to determine what is being typed. You would be wrong. This program uses the accelerometer in the phone in order to determine what keys are being pressed with up to 80% accuracy. The accelerometer only needs to be as accurate as an iPhone 4 or the equivalent. No one thinks that the tilt is a backdoor or a possible security problem that could be used against their security, which makes this program simply brilliant.

Cellphone Sniffer

Ok, maybe that's not what this is actually called, but that is an excellent description for this next cellphone hack. You know all of those fancy, ease of use credit cards that don't require contact to work? With these credit cards, all someone has to do is wave their credit card to complete a purchase. That's pretty amazing, especially for you. Get yourself a phone with NFC, or near field communication, then download a specific scanning program. From there, you move into the close proximity to get your phone within range of the phone's radio chip, which is designed to be scanned from a few inches away. Now, you can be a pick pocketing hacker. Thank Mr. Eddie Lee for this. He did release the information in very public arenas in an attempt to get credit card companies to fix this, but until then, thank you sir!

Juice Jacking

You know how your heart rate rises and you have a mini panic attack every time your cellphone is about to die and you don't have USB outlet nearby? You aren't the only one. People are notorious for letting their phones run near dry and waiting until alerted before attempting to charge again. How can you take advantage of this?

Well, use what you learned from Shoulder Surfing once again and don a disguise.

This time you are going to want to go undercover as some kind of technician. An electrical technician or a network engineer would be perfect.

So here is the setup. You need to locate or "provide" a free charging station. Using your disguise to access or set up this charging station, configure it so that you can steal personal data, passwords, and selfies. You can also upload some of your own programs that can benefit you in the future. Now you know why you shouldn't charge your phone at Def Con!

Piggybacking

Everyone loves wifi, no one more than hackers. That is why you should absolutely fall in love with a cell phone that is setup as a wifi hotspot. You can use this as an untraceable opportunity. Let's say that you have acquired some data that you want to sell, but you don't want it to lead back to you in anyway. Well, you can conduct your crafty entrepreneurial business using that hotspot, so that if the authorities do come looking, the find someone else.

Chapter 14: Understanding Computer Viruses

You might have heard of malicious programs attacking computer systems across the world and compromising their data security. Well, these programs are basically designed by hackers to access remote systems or steal their data, or maybe just for keeping an eye on them.

'Viruses' is a general term used by the media as well as the laymen for these malicious programs. While viruses are a major part of them, the whole segment of these programs comprises of other malicious programs as well. Let us take a look at them one by one:

Trojans

A Trojan, or a Trojan horse, is a program that gains entry into a computer system and allows the hacker to take control of that system remotely.

For most of the people who do not know, the name Trojan Horse comes from the ancient Trojan War where the Greeks used a wooden horse to gain entry into the walled city of Troy. This horse was called Trojan Horse and led to the destruction of the entire city from the inside.

Similarly, a Trojan horse is something that you should be scared of. The most dangerous thing about Trojan horse is that it is hard to detect when it is entering a system. The Trojan horse usually appears as something that is vital for the computer system, like an important program or a driver.

However, once a Trojan horse is installed, your system is compromised. A hacker can do anything with the help of a Trojan horse. It could be installing backdoors into your system, or download any data or malicious programs that the hacker wants to download in your system.

Some of the major tasks that can be accomplished (and are commonly accomplished) using Trojan horse are:

- Damaging the computer system. The effects of this could be repeated crashes of the system, blank display with just a blue screen, freezing, etc.

- Stealing of confidential data can also be accomplished using Trojan horse. This data could be passwords, bank information, credit card details, etc.

- Trojan horse can also modify or erase the files that are present in the system, resulting in loss of important data.

- Using the Trojan horse, a hacker can make the system a part of a DDoS attack on a web server.

- Trojan horse can also help the hacker in stealing the money of the owner by making bank transfers through his system.

- Using this malicious program, a hacker can make a note of all the keys that the user is pressing. This data can enable him to attain user ids, passwords, and other personal details of the user that are not meant to be shared.

- The hacker can view the screen of the user, as well as screenshot his screen to keep an eye on him.

- A hacker can download the history of the web browser of the user, thereby knowing the websites the user has visited.

Worms

Worms are very harmful programs like the Trojan horses. However, they function differently. They propagate through the system and try to replicate themselves as much as they can. They even find ways to get copied from one system to another system in a network.

By using worms, a hacker can accomplish the following tasks:

- The primary purpose of worms is to install backdoors on the target systems. These backdoors are like an entryway into the computer system. They can be used to do distributed denial of service attacks on the servers. These can also be used to send spam emails from the computer system and all the systems that it affects.

- Worms keep on propagating by copying themselves. By doing this, they tend to slow down the entire network by taking a lot of the bandwidth for themselves.

- Worms can also install various malicious programs into the computer, like other worms, Trojan horses, etc.

Viruses

Viruses are malicious programs that get attached to the various programs and files that are used by the target system. They keep on running all the time and hence consume a lot of memory of the computer, as well as its CPU usage. The programs or files that the virus attaches it to are termed as 'infected.'

It works in a similar fashion as the biological viruses, which tend to get attached to the normal cells of the human body and eventually alter the body's function. As a result, people tend to fall sick. Similarly, a computer virus can cause the sickness of a computer.

The viruses not only hide well, but they would appear so alluring that you, as a user, would be tempted to open them. It might be a document in your mailbox that seems important or essential data on your flash drive. Once the infected program is run, the virus code starts up.

With a virus, a hacker can accomplish the following tasks:

- Getting access to the confidential user data such as his user ids and passwords.

- Destroying the data present in the system

- Corrupting the system data and drives

- Send messages or advertisements to the user

- Keep a log of the keys that the user presses

How to Combat These Malicious Programs

As an ethical hacker, not only is it important to understand about these programs, but you should also know how to tackle them. The following measures are essential in dealing with these programs:

The primary step to follow is to install a good anti-virus software on the computer system. Anti-virus checks all the potential threats automatically and protects the computer system from most vulnerabilities.

Organizations often create policies that disallow the users to download or visit websites that might download any unneeded files from the Internet. These could be any games, toolbars, email attachments, or any programs that are not required.

Any flash drives or secondary storage device that is ever attached to the computer system should be scanned nicely. External storage devices are a common way of propagating the malicious programs. Hence, it should be made sure they don't have a virus before you open them.

It is a good idea to back up important data. Even if there is an attack of a malicious program that corrupts the file system, the backup will ensure that data is not lost. This backup should be done on read-only devices, such as a CD or a DVD, so that virus cannot affect these drives.

It is always a good idea to keep your system updated to the latest version. Most malicious programs tend to exploit the system vulnerabilities, therefore keeping your system updates makes sure that the vulnerabilities are not present. Companies roll out updates with the same motive.

All email attachments should be checked before opening or downloading. It is a good idea to download attachments from known sources only. And if something seems important enough to download, remember to scan it with your anti-virus software.

Chapter 15: Hacking The Network

What good is hacking without the Internet? There are a lot of people that want the skills of hacking just so that they know how to access secured networks and reap the free wifi benefits. Whether it is to get free wifi, snoop on your neighbors, or other clandestine activities, you need the knowlege of hacking strategies to hack into a network.

Get That Wifi

This might be obvious, but it needs to be said. Before you can actually connect to the network, you first have to be in range. It's not very fruitful to attempt to hack into a network that you can't even connect to. So, once you are in range, the games can begin.

First, open up your trusty BackTrack and crack that wifi using AirCrack-ng. There are several tutorials available for the use of both programs to help you if you get lost in this process.

To start, you should place your wireless card into monitor mode. You can move the wireless card into the mode with:

bt > airmon-ng start wlan0.

After this, start airodump-ng, which allows you to dump info on all available devices to your monitor. That is:

bt > airodump-ng mon0.

What you see is determined upon the strength of your wireless card. If you have access to a high gain antenna, your results are going to be much better. The names of the networks will be listed, as well as the strength of the signal, as indicated by PWR, which is the second column. So, you want to choose one of these to continue. If you have a particular target in mind, you may want to do a little reconnaissance to determine exactly which network you are wanting.

Now you are going to want to capture the password hash for this network. To do this, you first have to force whoever is present off of their access point. You do this with a deauthenticate, or deauth command, sent with airoreplay-ng command. It may take a while, but you have that virtue known as patience.

OTHER WAYS TO GET IN

SneakerNet

The easiest way to access a vulnerable location is to simply walk through the welcoming and open front door, much like Shoulder Surfing. That is the concept behind this process.

There is an extremely simple way to get into a network, and it slips a lot of people's minds when they have hacking on the brain. If you can come up with a reason to, or can do so without being seen, just connect directly to the router with an Ethernet cable. Yeah, it's that simple. You immediately connect without having to enter a password. Not only that, you can access the router's settings. Even more so than Internet security, most people never adjust the router admin settings. Just like before, enter the default username and password. Once in, there are several open places that you can hack. You can change or view the wireless password and manage or disrupt saved connections. One other great idea would be to add your own MAC address to the router's whitelist. This will let you conveniently connect with no future issues. If the router has Wifi Protected

Setup, then getting on the whitelist is even easier. Simply attempt to join the network, then immediately press and hold the router's access button until you are connected. Presto chango, hack complete.

Wardriving

This one might not be for you, but who knows, you might need some fresh air. This method consists of driving around looking for unsecured networks. If you aren't finding many, than just look for WEP protected networks. WEP is relatively easy to crack in a minimal amount of time.

There is a downside, thought. The tools used to crack this security come from people that might want to hack you. So, unless you completely trust your source, do this.

Use PHLAK. It is designed to test a network's durability. Its lightweight nature allows you to run the program from a CD or flashdrive, keeping your hard drive safe, and protecting you and your information.

Once you have it mounted on your flash drive, simply reboot into the temporary OS.

Grrrrr…

Ok, so none of that worked. Well, you know what we have to resort to now. That's right, brute force! We have already gone into an in depth description on how that works, just know that it is an option for you here, as well.

You're In, What's Next?

So you got into the wifi. What now? Well, why not run an ARP protocol and enumerate all systems on the network?

You can do this to find out what all systems are available in the location, providing you a more comprehensive list of potential exploits.

Maybe you are a bit of a prankster, and you want to print something... unexpected with the printer connected to the network.

When you see a command, like below, substitute “IPaddress” with the IPaddress of the device you are looking into. Enumerate the systems with:

Bt > netdiscover -r IPaddress

Use Your Tools

Remember NMap? Let’s put it to use before you do anything. This will inform you of the open ports you can exploit.

bt > nmap -sT IPaddress

By understanding how IP address work, you can look through the provided list and determine what kind of devices are connected to the network. You may have to do a little bit of research along the way, but you already have most of the knowledge that you need.

Operating Systems

So you have a list of devices connected to the network. Now you need to find something vulnerable and susceptible to your exploitation. You are going to want to locate something Linux or Windows based that you can use your tools on. If you go for Windows, which is likely the kind of computer you are to find, you may be able to turn on a webcam and see what is going on. To start, you are going to want to scan the IP addresses that you suspect you can exploit. Do this with:

```
bt > xprobe2 IPaddress.
```

An essential thing to remember is that xprobe2 has been known to mistake Windows Vista and Windows XP sp2. If you get one of these results, assume that the operating system is Windows Vista. The reason for this is that Vista is more secure, and you know that what will work on Vista will work on XP. Following this process can half the amount of trouble you take trying to hack.

Perform the Hack

It is up to you to choose which hack to use at this point. Simply find an open port, and apply a hack to that vulnerability, if one exists. For example, if you find the port 445 open (working here in Windows Vista) there might be a vulnerability that you can exploit.

For this example, reach into your bag of tricks and bring out Metasploit and try a bit of code.

If you don't have it, download meterpreter. Once you get in, you can use this to turn on the webcam for some true reality TV. Next, set RHOST (your target) to LHOST (you). You may have to try a few times before meterpreter prompts properly. Now you are in. That feeling of power, it's pretty amazing isn't it?

Turn on That WebCam

Meterpreter has a script that you can run to turn on the webcam. First, disable the antivirus on the target computer, just to be safe:

meterpreter > run killav.rb

Now take a quick picture from the webcam, you don't know if you are prepared for what is on the other side just yet:

meterpreter > webcam_snap

Now, go pop some popcorn, put on some PJs, then head to /root/directory to view the .jpg.

There's no telling what you might see. Prepare yourself!

Chapter 16: Web Hacking, Xss, And Sql Injection

Over the past decade, the World Wide Web has expanded rapidly and is still evolving today. Traditional systems are being replaced by dynamic, browsable applications that are hosted on web servers and access the vast database. The continued implementation of the broadband Internet has paved the way for multimedia enhancements. And the dramatic evolution of wireless technologies today offers WWW a chance to reach anywhere.

Undoubtedly, the benefits of the World Wide Web are to overcome spatial and temporal limits. Today, we can order from the comfort of your home a new phone that has just been produced in Japan, order it through the web, or apply for a job remotely in South America with one click. In business, the Web is being exploited more and more, as it makes access to customers and opportunities much easier. The bank services are now online. Shopping goes on the net with fast and easy payment by credit cards. And more and more e-business companies are born.

But for any web-based application that goes online, and any e-business that powers up a rack of servers, malicious hackers with appropriate attack techniques are also available. While people are coming to terms with new technologies and making their network, web servers with firewalls, and defense systems more secure, the attackers have also become smarter so they can break these systems and firewalls. The continued rapid growth of web technologies also leaves many security gaps. In this book, as we mentioned before, the well-known web-application-hacking techniques are presented, and avoidance methods are shown.

First, let us discuss an overview of web application architecture.

Web Application Architecture

Traditional HTML pages are static, that is, they will eventually be generated and will be available in that form on the Web server through the URL of the page. Of course, it is also desirable to have an HTML page dynamically generated, for example, to influence recent page data of the caller. Dynamic web pages are generated by a web application at runtime.

Web applications generally differ in client-side and server-side web applications.

Client-side web applications are run on client machines. The known technologies for such web applications are JavaScript, VBScript Flash, Applet. Server-side web applications are executed on the server, often used for the dynamic generation of data, e.g., from a database. The well-known technologies for server-side web applications are Common Gateway Interface (CGI), Server APIs, and scripting languages such as ASP, PHP, JSP.

A common dynamic web system consists of 4 components: the web client (often a browser), the front-end web server, the server applications, and the server database. The front-end web server acts as an interface for interfacing with the outside world, getting input from the web client via HTML forms and HTTP, and returns an output generated by the application in the form of HTML pages.

The server application works with the database to perform transactions.

Each component of the web system has its own specific vulnerabilities. Here is a short overview of the basic types of attacks on each component:

· Web client: execute active content, client software vulnerability exploits, cross-site scripting (XSS).

· Web server: server software security vulnerabilities exploitation.

· Web application: an attack against authentication, authorization, input validation.

· Database: execute privileged commands through database query, manipulate queries (SQL Injection) to get excessive record.

Of these, XSS and SQL Injection have been selected as web-based tools.

XSS

Cross-site scripting (XSS) exploits a security gap in the attacked application and can be used to modify the data of an original page, thereby paying for the active attacks. In XSS, information is embedded by an attacker in a supposedly secure site. [ScSh02]

One can imagine that a malicious user leaves a few lines of JavaScript code as a message in one guestbook for others. In the event that the guestbook does not consider this user's input suspect, script code will be embedded in the page and later exported to any user's browser that allows JavaScript code. Often important information such as access data, personal or financial data is stolen in the cookie.

The easiest way to test an input field if it contains an XSS security gap is to type the following line:

```
<script Language = "Javascript"> alert ( 'hello '); </ script>
```

If the browser opens a window with the text 'Hello,' this means that JavaScript has been written to the guestbook as script code. Then the user browser reads the HTML page with the written script code and executes it.

Types of XSS

Depending on location (client-side or server-side) and duration (permanent or non-permanent), cross-site scripting attacks are divided into three basic types.

Type 0

Type 0 are the XSS latches that contain the manipulated script code on the client-side.

For example, this could be a JavaScript code that gets a URL argument value and writes unfiltered:

http://www.example.com/search.cgi?query=<script>alert(document.cookie); </ Script>

When the client browser submits this URL, it then gets an HTML page containing the JavaScript code. This embedded JavaScript code is executed on client machines with current client rights to steal cookies. The code in < script > tag is unlimited, and it can also be replaced by a script code from other servers.

Example scenario:

Is there a link?

http://www.example.com/search.cgi?query=<script>alert(document.cookie); </ Script> published in the forum or on a web page.

Alice clicks on the link

The malicious script opens an HTML page on Alice's computer

This HTML page contains the malicious script that is executed on Alice's computer with its privilege rights.

Type 1

In this type, the manipulated script is on the server-side and not permanent. There is no problem with the subsequent execution of the script with non-random input. For example, a search function on the server that gets the search term ff as input without filtering and outputs something like this:

The one you searched for: "search term" is in .....

And she is using search string ff = < script > alert (document.cookie); < / script > , which provide output:

The one you are looking for: <script> alert (document.cookie); </ script> is located in .....

The script code is executed again on user computers in order to steal cookies with sensitive information.

Example scenario:

Alice often visits a Web site A that contains XSS security holes.

Tom creates a manipulated URL to site A and sends it to Alice

Alice visits the manipulated URL

The embedded script in the URL will be executed in Alice's browser as if it came directly from web page A. The script can steal sensitive information (banking account, email account, ...) and send it to Tom without Alice's knowledge.

Type 2

With XSS with Type2, the manipulated script code is permanently written to a web page. It is the most dangerous form since it affects many users. This type 2 works similarly like Type 1, only that the output persistently stays on the page.

Example:

Website A, again containing XSS security holes, allows users to post news or other content.

Tom posts a message on this website:

"A really good website! <Script> alert (document.cookie); </ script>”

When looking at the news, the script code is run on user machines and therefore can user cookies, other things without his knowledge to Tom.

Protection

Website operators should never trust user input. All incoming input values must be viewed and filtered. You should define an input table exactly and only allow such input.

XSS attack actually embeds a manipulated script code between 2 < script > tag and < / script > tag (or < object>, < applet>, < embeded >). If these tags are removed, the malicious script code is no longer executable. You can filter these tags by typing through, for example, the following PHP code:

```
function filter ($ input) {

$ input = ereg_replace ("<script>", "", $
input); $ input = ereg_replace ("</
script>", "", $ input); $ input =
```

```
ereg_replace ("<object>", "", $ input); $ input = ereg_replace ("</ objectt>", "", $ input); $ input = ereg_replace ("<apllet>", "", $ input); $ input = ereg_replace ("</ applet>", "", $ input); $ input = ereg_replace ("<embeded>", "", $ input); $ input = ereg_replace ("</ embeded>", "", $ input);

return $ input; }
```

This removes all entries with dangerous tags, e.g., the above cookie stealing. Script code looks like this after filtering:

```
alert (document.cookie);
```

It is, of course, no longer harmful! In some cases, the input should be clearly printed unchanged. You can do that by entering the user escapes, that is, all special characters are with replaced their equivalent in HTML- the so-called escape sequences. These escape sequences are a sequence of normal characters that represent their special characters. For example, the string < script > will appear as a normal string < script > . displayed later as a string < script > by the browser, which is recognized as no more tag. The conversion of special characters can be realized with the function htmlentities ().

```
function escaping ($ input) {

$ input = htmlentities ($ input);

return $ input;

}
```

This method is a simple and secure way to deal with XSS. Nevertheless, there is a big disadvantage that all special characters and thus, all tags are blocked. Users can, of course, not use HTML tags to format their text.

By turning off ActiveScripting in the browser, you can protect against XSS on the client-side - no manipulated script code is executed on the client-side. Nevertheless, this does not help with pure HTML injection (e.g., with < iframe > tag ...), but it is not dangerous like real XSS. It cannot steal cookies from clients under any circumstances.

Conclusion

I hope you learned a lot! I hope that this book helped you gain useful knowledge on how to hack computers and networks ethically, and I hope that you learned how to protect your system from future attacks as well.

The next step you should take is to take all of the things you learned and put them into practice. The more you use your newfound knowledge, the more familiar you will be regarding the intricacies of computer networks.

Yes, being hacked is a scary possibility, but with a bit of knowledge you can prevent unethical hackers from getting inside your system.

Thank you and good luck